William G. Mather

Great Lakes Steamship

By Thomas G. Matowitz Jr.

Photo Museum Guide

Dedication

To the Great Lakes sailors whose skill saw the *William G. Mather* safely through 55 shipping seasons and to those who came after: the staff, volunteers, and friends whose hard work has assured the ship's survival.

For John Gribben

The *William G. Mather* Museum would like to thank the following people whose generous financial support helped make this museum guide possible:
The Propeller Club - Port of Cleveland
John L.Horton
Anonymous Donor in the appreciation for the 250,000 hours of "sweat equity" to restore, maintain and preserve *"The Ship That Built Cleveland"*

Text by Thomas G. Matowitz Jr.
Editor and design Randall Shoker *Typist* Karen Fergus

ISBN# 1-930127-03-0
Published by the Oxford Museum Press, Inc., Oxford, Ohio Visit us at www.oxfordmuseumpress.com Printed in the USA

Captain Harry Anderson during his tenure as the *William G. Mather*'s commander in 1967. At the time of this book, he is 91 and is still active in the affairs of the ship.

Contents

In a scene familiar to generations of Clevelanders, nearly a dozen freighters wait to unload or take on cargo. This view from October, 1949, shows the old Cleveland Stadium and a very different skyline.

Forward

The idea of establishing a Great Lakes history and shipping museum in Cleveland, Ohio, had been a persistent goal among Northeast Ohio maritime enthusiasts since the 1970s. The realization of the Cleveland dream began in December 1987, when the Cleveland-Cliffs Company donated the *William G. Mather* to the Great Lakes Historical Society of Vermilion, Ohio, for preservation. The former flagship *Mather* is a most appropriate symbol of Cleveland's extensive involvement in the Great Lakes industry during the 20th Century's golden era of steamboats. She had been named for a well-known Cleveland civic leader, was a fixture on Cleveland's waterfront between its runs to the Upper Lakes, and was the site of several technological innovations during its remarkable 55-year career.

The *Mather* arrived in Cleveland in 1988 to begin its transformation. Volunteers toiled alongside professionals as the *Mather* was restored and adapted for its new career as a floating maritime museum. The *Mather* Museum opened in 1991 as the first "new" attraction in Downtown Cleveland's developing North Coast Harbor Park. However, the lack of a long-term lease for the *Mather*'s East 9th Street Pier berth threatened to close the Museum forever in 1994. Fortunately a group of concerned individuals banded together in 1995 to save the *Mather* Museum. That group, the Harbor Heritage Society, is now an Ohio nonprofit corporation dedicated to the development of the 618-foot *Mather* as a premiere maritime attraction. As of this spring 2001 writing, the Steamship *William G. Mather* Museum has truly become a premiere educational maritime site. The Greater Cleveland community has contributed more the $6 million dollars of resources to this effort, including the donation of the ship itself and more than 250,000 volunteer hours.

The *Mather* is a work in progress. Over the next several years, the Museum plans to expand to a year-round facility and undergo a major expansion of its onboard exhibitions. I invite you to learn about the historic *Mather* as you come aboard to experience this evolving Great Lakes treasure firsthand.

James D. Ireland III *President, Harbor Heritage Society*

A tug from Great Lakes Towing assists as the *William G. Mather* is moved to its East 9th Street Pier location.

Introduction

As one of the few surviving straight-deck bulk freighters, the *William G. Mather* represents a vital technology that, once commonplace on the Great Lakes, has practically disappeared.

The familiar straight deck bulk freighter design began with the steamer *R. J. Hackett* launched in 1869. This ship was the first to be built with engine and boilers located aft and the pilothouse forward, providing an unobstructed main deck for ease in loading and unloading cargo. Improved shipbuilding methods, and the use of iron, and ultimately steel, led to the construction of six hundred-foot bulk freighters by the early 1900's. These ships were well suited to the task of hauling thousands of tons of bulk cargo, typically iron ore, coal, grain, and limestone. While they were easily loaded, the question of how to unload these large ships efficiently was not fully resolved until the invention of the Hulett unloader. These shore facilities were critical to the success of the straight deckers. A mechanical clamshell bucket device whose operator actually rode inside it, these machines could unload a vessel like the *Mather* in several hours.

At her launch in 1925, the *William G. Mather* was one of the largest bulk freighters on the Great Lakes. Originally powered by a quadruple expansion reciprocating steam engine, and notable for her elaborate guest quarters, she represented the highest development of modern shipbuilding at the time. The *Mather* and her sister ships played a vital role in our nation's economy. During its long working life, the *Mather*'s primary task was to haul iron ore from western Lake Superior to ports on the south shore of Lake Erie to serve as raw material for the steel mills throughout the Midwest. This trip took the *Mather* a week to complete. It was two and a half days upbound, a day to turn around, and two and a half days downbound.

This view looking toward the stern of the *Mather* gives one a good feel of the ship's 618 foot length

One of the pair of winches used to open and close the cargo hatch covers.

The round trip was over 1,600 miles and was completed more than thirty times in an average year. The ship performed this routine faithfully for more than half a century.

By 1980, time and significant changes in ship design were about to overtake the *Mather*. One of the largest ships on the lakes at her launch in 1925, she was dwarfed by freighters four hundred feet longer and nearly twice her width. It was not simply a matter of size. The thousand footers were self-unloaders, meaning that they had a system of conveyors and booms that enabled them to discharge cargo at any port they could reach. No longer were bulk freighters limited to destinations where Hulett unloaders were available. This meant that an aging straight-decker could no longer operate profitably, and so it was that the *Mather* made her last commercial trip in the autumn of 1980.

She was eventually laid up in the Maumee River at Toledo, Ohio, in an area known as the Frog Pond. Her future did not look bright as the early 1980's saw dozens of veteran Great Lakes cargo vessels sold overseas for scrapping. The *Mather's* escape from the scrapper's torch was a narrow one, but on December 9, 1987, the Cleveland-Cliffs Company gave the ship to the Great Lakes Historical Society to be restored as a museum for educational purposes. This is when a band of truly dedicated people found their work cut out for them. Ships in active use are subject to frequent, if not daily maintenance. The absence of this leads to rapid deterioration. In the case of the *William G. Mather*, it was more than benign neglect. A fire ravaged the aft deckhouse, virtually destroying the guest's and officer's dining rooms, and causing major smoke damage elsewhere. Unwilling to give up in the face of such obstacles, the restorers began a project that continues today: returning the ship to the impressive condition of a Cleveland-Cliffs flagship that, serving as a museum, has educated and entertained hundreds of thousands of visitors since opening in May 1991.

A scuttle on the main deck to provide access to the cargo hold.

Construction and Launch

In the spring of 1924, officials of Cleveland-Cliffs contacted the Great Lakes Engineering Works at Detroit, Michigan, to place an order for a new ship. Assigned constructor's hull number 250, it would be christened *William G. Mather* and become the new Cleveland-Cliffs flagship.

The ship's design had several novel features. The ballast tanks lining both sides of the cargo holds had sloping tops. This prevented cargo from being trapped on top of the tanks, making unloading easier. The ship's cargo hold was divided by three screen bulkheads. This added great strength to the hull and minimized damage done by unloading machines. Another innovation was the design of the vessel's pilothouse. Its wedge-like floor plan gave officers a clear view of the bow and sides of the ship. This provided excellent lines of sight for the ship's officers during docking and maneuvering without requiring them to leave the pilothouse in bad weather conditions.

Building took a year, and the cost of construction was nearly one million dollars. At noon on Saturday, May 23, 1925, the *William G. Mather* was launched. She was christened by Ms. Clare Schneider Kirchner, daughter of the manager of the Cleveland-Cliffs Marine Department. It is interesting to note 66 years later to the day, Mrs. Kirchner participated in the ship's dedication as a museum.

A ship's launch is certainly an important day, but much remains to be done before it is ready to enter service. In the case of the *William G. Mather*, this process took almost exactly two months. On July 24, 1925, commanded by veteran Cleveland-Cliffs Captain Charles A. Anderson, she departed the shipyard to begin her first trip. The *Mather*'s destination was Sandusky, Ohio, where she took on a cargo of coal destined for the Canadian port of Fort William, Ontario. Her 55-year career as a bulk freighter was underway.

Left: This unusual view shows the *Mather* from dead astern just prior to her launch. Notice the large single four bladed propeller as well as the rudder and the crude wooden walkway leading up to the ship.

Right: The bow of the *Mather* as she waits to take to the water. Notice the box like shape of her hull which made her an efficient bulk carrier.

Riveters working in the cargo hold of a bulk freighter under construction, circa 1925. It is a good illustration of the equipment and methods in use at the time. Note the worker on the scaffold preparing to catch a white-hot rivet thrown to him by the man standing in the foreground.

These dignitaries took part in the *Mather*'s launch. The young lady with the bouquet of flowers is Clare Schneider Kirchner, who christened the ship. To her immediate left is Mr. William G. Mather.

The *Mather*'s launch in progress. This type of launch was common on the Great Lakes where shipyards were often located on waterways too narrow to permit a conventional stern-first launch.

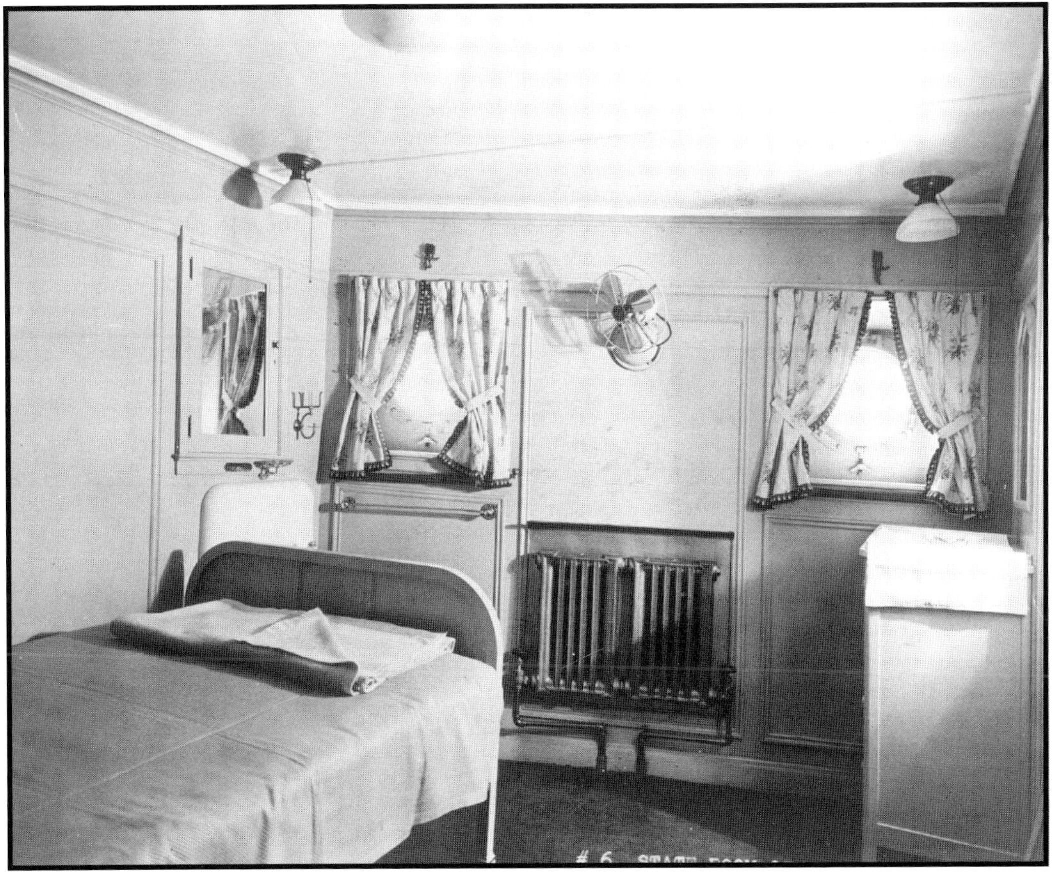

One of the *Mather*'s guest rooms as it appeared during the 1950's.

William G Mather, flagship of the Cleveland-Cliffs fleet. Here is an early photo with her original funnel.

Restoration

The *William G. Mather*'s restoration has been a lengthy process which continues to this day.

When the ship was retired in 1980, its days seemed numbered. While it was moored near Toledo waiting for its fate to be determined, dozens of its contemporaries were sold for a fraction of their original cost and cut up for scrap. While the *Mather* languished, something happened which should have sealed its fate. A vagrant broke into the ship's aft deckhouse and started a fire. Centered in the guests' and officers' dining rooms, this fire also extensively damaged adjoining areas. This, of course, made the ship's return to service even more unlikely.

By 1987, very few ships of her type still existed, and Cleveland-Cliffs reached the decision to donate the *Mather* to the Great Lakes Historical Society in order to serve the public as a museum. Because of the ship's long association with Cleveland, it was decided this would be the best location for the proposed museum.

After seven years of inactivity, it was not possible for the *Mather* to move under its own power, so it arrived in Cleveland with the aid of tugs at dusk on an autumn evening in 1988. It was intended from the start that the *Mather* museum's permanent location would be on the lakefront in downtown Cleveland's North Coast Harbor Park. However, while the major tasks of restoration were in progress, it was necessary for the ship to be located elsewhere. This alternate location proved to be a pair of docks located at Collision Bend on the Cuyahoga River's east bank.

Work could begin in earnest. The first step was to prepare a detailed plan to determine how the proposed museum could best serve as an educational resource. It was also vital to realistically estimate the financial cost of the ship's physical restoration and plan the necessary fundraising activities. Even apparently simple problems, such as the method of mooring the ship at its proposed East 9th Street location, required careful thought. Rough weather, particularly during the winter months, is entirely capable of breaking conventional mooring lines. A new system had to be designed. An engineering firm accomplished this goal, and at a cost of more than $100,000 the problem of how to moor the ship securely year round was solved. With all this preliminary work done, the actual restoration of the ship itself began in 1989. Volunteers began spending Saturdays performing many of the tasks required to undo nearly a decade of neglect, and to overcome the serious obstacles created by the fire in the dining rooms. Sandblasting was a major task during this time as multiple coats of paint dating back to 1925 were removed so surfaces could be prepared for fresh paint.

In October 1990, the ship was ready to be moved to her new location. With the assistance of tugs, this was accomplished safely, and the final tasks necessary for the *Mather* to be opened to the public were undertaken. On May 23, 1991, the 66th anniversary of the ship's launch, the *William G. Mather* was dedicated as a museum. An honored guest was Mrs. Clare Schneider Kirchner, who as a young woman many years earlier christened the ship on her launching day.

Left: Clare Schneider Kirchner at the *Mather*'s launch in 1925.

Right: Clare Schneider Kirchner, who christened the *Mather* in 1925, is joined by family and friends on the day of the ship's dedication as a museum, May 23, 1991.

The Ship Today

Open to the public at its permanent berth at Cleveland's East 9th Street Pier since May 1991, the *William G. Mather* has informed and educated hundreds of thousands of visitors. Many are school children seeing such a ship for the first time. Some are former crewmembers who traveled across the country to show a grandchild the ship that was once their home. Many visitors remark that they grew up seeing such vessels frequently but were never permitted on board because of the owners' concerns about safety and liability.

A visit to the *Mather* today normally includes a tour led by a well-informed guide who is quite possibly a former Great Lakes sailor. Tours last about an hour and take the visitors through the whole ship from the dunnage room to the engine room. Thanks to a great deal of hard work, the ship looks much as it did during its working life. Current visitors may well be greeted by one of the ship's former captains, brisk and active in his nineties, and should not be surprised to find a retired steward preparing food in the ship's galley.

Educational programs for children are numerous, and summer evenings often provide a speaker on varied subjects of nautical interest. A summer tradition is the whistle pull held on the 4th of July. A large group of steam whistles, many of them from well-known and long-vanished Great Lakes ships, may be heard once more. Over Labor Day weekend, guests find the *Mather* an excellent vantage point from which to observe Cleveland's National Air Show. Future plans include an even wider range of programs to educate and entertain visitors who may shortly be able to visit the museum year round.

It is entirely possible that the *Mather*'s 55 years as an ore boat will prove to be simply preparation for her true career as a teacher.

Ship's equipment and machinery

The *Mather's* anchor windlass room. This view of the anchor windlass illustrates two safety devices. The black hook suspended by a chain from the overhead is a devil's claw. The hinged black bar attached to the mouth of the hawsepipe jams against a link in the anchor chain. Both prevent the stowed anchor from paying out accidentally.

Taken on the spar deck, this photo shows the telescoping hatch covers and the clamps that secure them.

Seen on the boatdeck, this is the engine room skylight. It is a source of natural light and ventilation for the engine room more than twenty feet below.

The steam turbine-powered generator that provides the ship's electrical power.

This is one of the ship's ballast pumps

This evaporator provided pure water for the boiler.

Located in the steering flat, this is the ship's steering engine. It provides the power to move the rudder in response to the ship's wheel.

Left: The added power provided by the steam turbine made a new propeller necessary. The five-bladed unit, seen here in drydock, replaced the original four-bladed propeller installed when the ship was new.

Below: One of two spare propeller blades stowed on the boat deck.

Shaft alley. Here is the propeller shaft looking aft.

William G. Mather

Machinery spaces

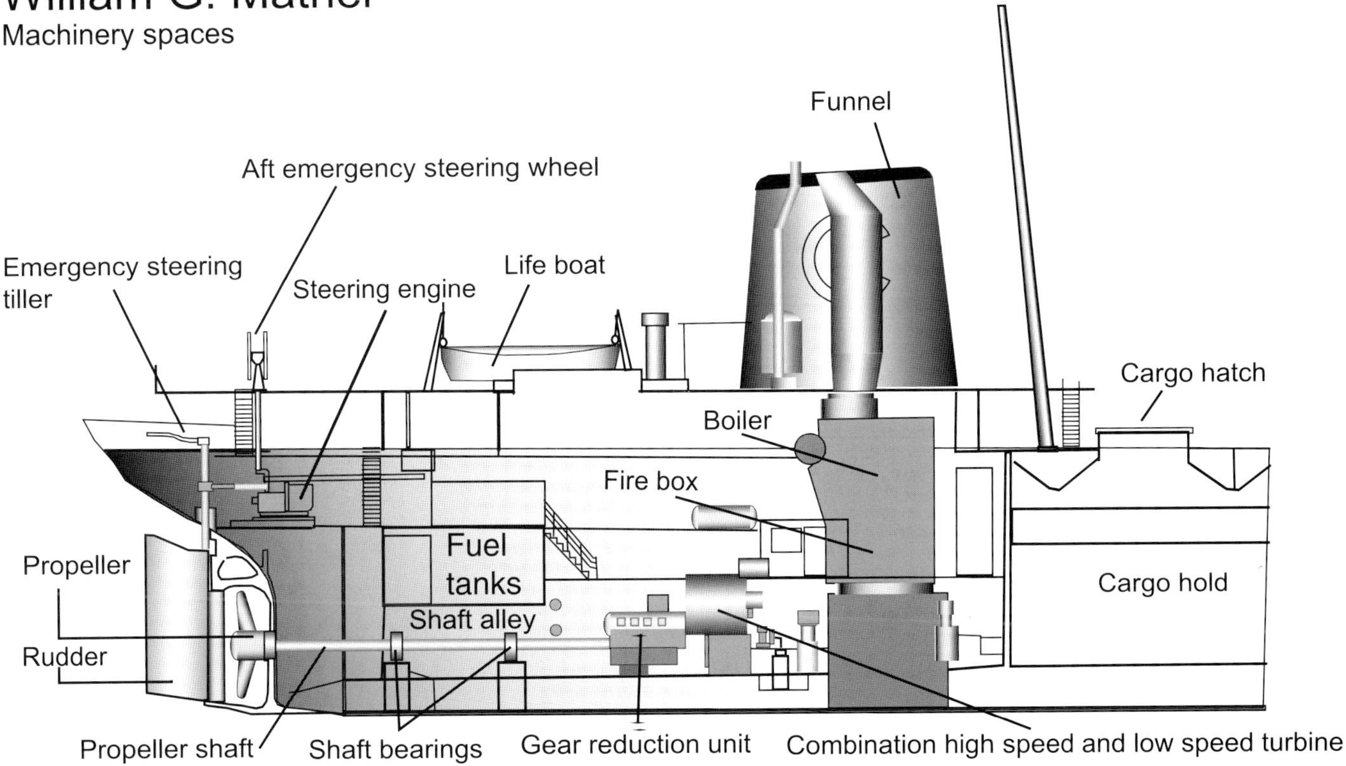

Drawing by Randall Shoker

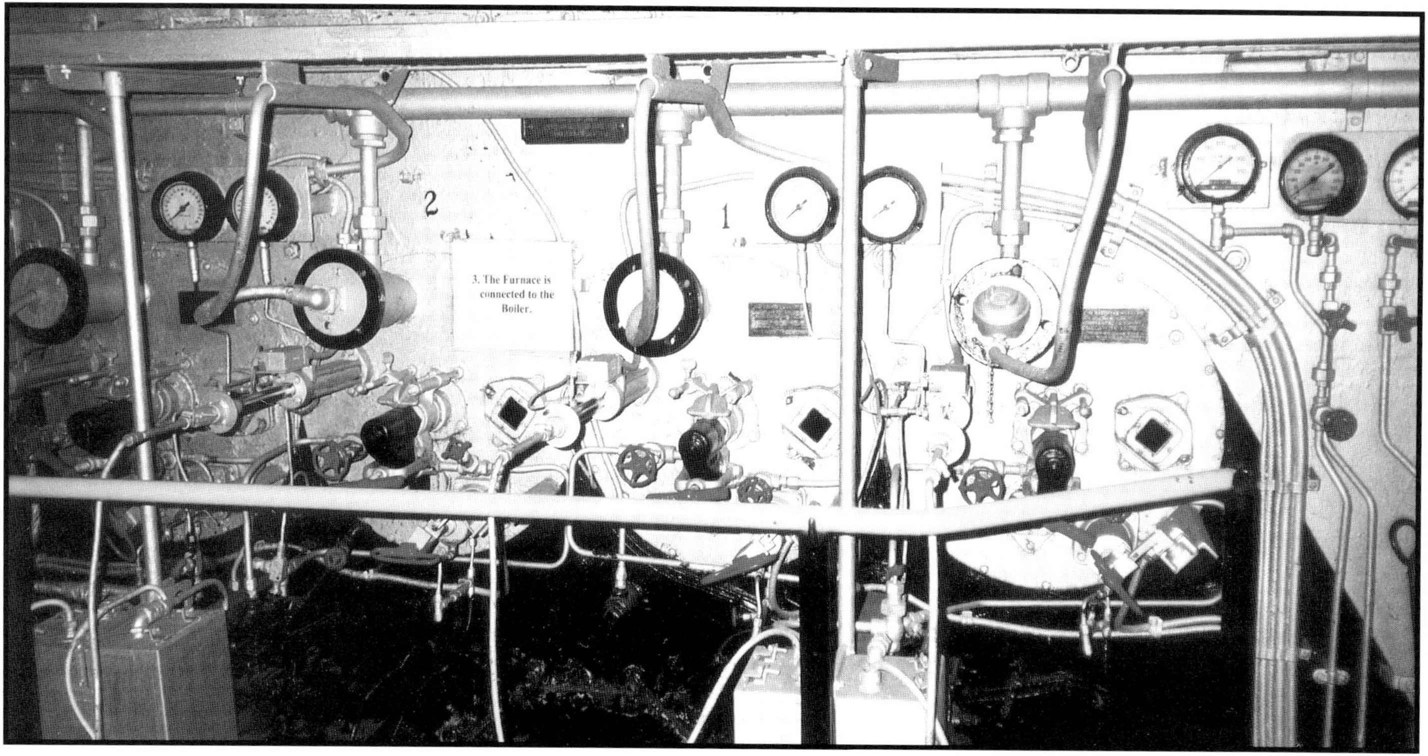

Taken from the watch engineer's station in the engine room, this view shows the plumbing and injectors that supply fuel oil to the burners that heat the boiler.

This view of the engine room shows the turbine in the foreground and the watch engineer's station beyond.

Seen from the spar deck, the forward deckhouse contained the pilothouse and accomodations for the crew, officers, captain and guests.

The Cleveland-Cliffs company flag. Its design changed over the years with the above flag being the last used by the *Mather*.

Both the old and new funnels were painted black with a large scarlet C. The C was made of of steel that was welded on.

The company name was often displayed in large plain letters amidships.

Ship's Markings

Drawings by Randall Shoker

Beyond the ship's wheel, this view shows, left to right, a radar unit, the gyrocompass repeater, the binnacle and the second radar. Partially visible on the extreme right is the radio direction finder.

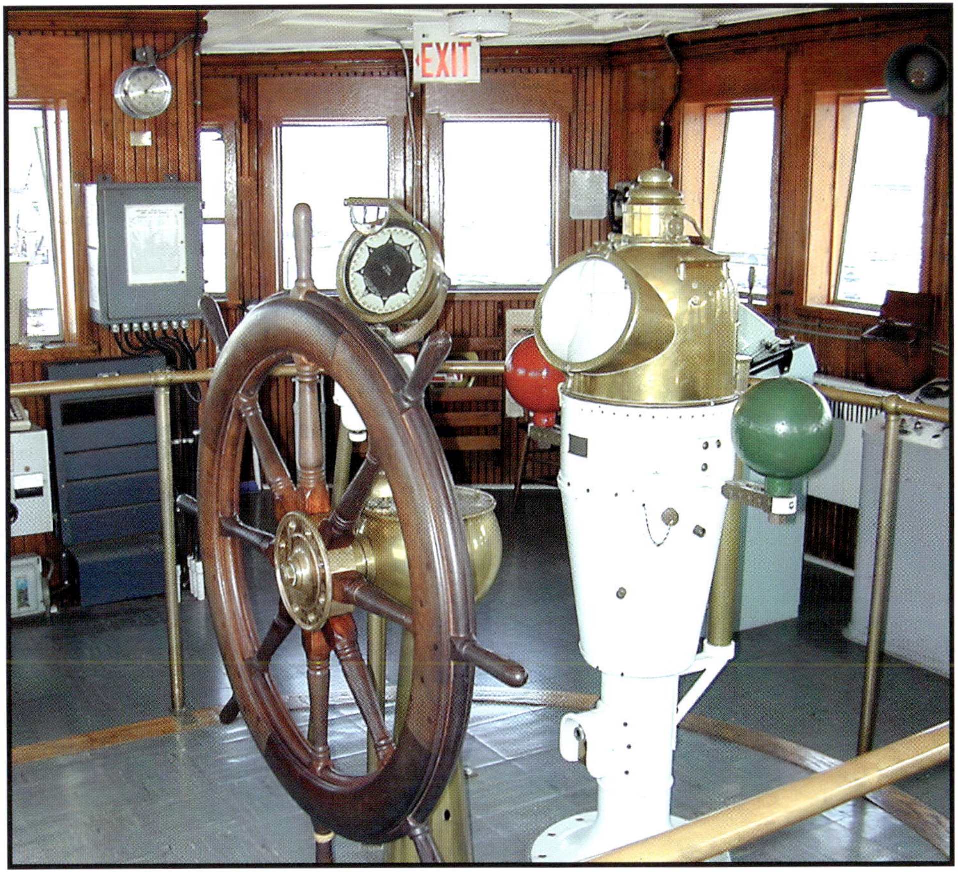

The wheelsman's post. Directly in front of the wheel is the binnacle containing the magnetic compass. Left of the binnacle stands the gyrocompass repeater.

The gyrocompass repeater relays the heading shown by the directional gyro located in the dunnage room.

The engine room telegraph. This device signals the engine speed & direction desired by the officer on watch in the pilothouse.

This radar set is one of two located in the pilothouse.

With the cover removed, the radar screen is revealed.

The radio direction finder located in the pilothouse.

The radar antennas on the pilothouse roof.

The *Mather's* bow. It was not uncommon to see sun chairs on this part of the ship for the *Mather's* guests. These are visible in the photo on page one.

The aft deckhouse is where the galley and engineering staff lived and worked. To the left of the funnel is the ship's boarding ladder in its stowed position.

Located aft on the boatdeck, this is the emergency steering wheel. It served as a backup in the event the steering engine had a mechanical breakdown.

Viewed from the dunnage room, this is living space occupied by deck hands.

This table in the crew's mess is where deckhands ate their meals.

The observation lounge in the guest quarters.

The guest dining room.

With the assistance of a tug, the *William G. Mather* moves to the East 9th Street Pier to begin her new career as a museum.

Working Life

Seen in October 1975, another sunset finds the *William G. Mather* still at work.

Working Life

The Beginning

Mr. William G. Mather and the Cleveland-Cliffs Company

In 1850, the Cleveland Iron Mining Company was formed. One of its founders was Samuel L. Mather, a prominent descendant of the notable family of New England clergymen. His son, William Gwinn Mather, born September 22, 1857, was prepared from an early age to follow his father into the iron ore business. After graduating from Trinity College in 1877, he joined his father's company, beginning his career as a clerk.

With his father's death in the summer of 1890, William G. Mather became his company's President at the age of 32. The next year, Mr. Mather and his associates acquired a controlling interest in a business rival, the Iron Cliffs Company. The result of this merger became one of the best-known names in Cleveland business – The Cleveland-Cliffs Iron Company. Headquartered in Cleveland to this day, it remains North America's largest iron ore mining company.

William G. Mather led the firm as President for more than forty years, serving in this capacity until 1933. From that year until 1947, he remained active in the company's affairs as Chairman of the Board of Directors. He retired in 1947 at the age of 90. Throughout his long life, he was very active in the Episcopal church and was deeply involved in Cleveland's business and cultural life. He served as a trustee of the Cleveland Museum of Art for thirty years and helped it to acquire many notable works of art. With good reason, he was known as Cleveland's First Citizen.

Mr. Mather was also interesting due to the circumstances of his marriage. Previously a confirmed bachelor, on May 18, 1929, at age 72, he married his neighbor, Elizabeth Ring Ireland, a 38-year-old widow with a young son. A devoted couple, Mr. and Mrs. Mather took their wedding trip on the steamship named for him. They remained happily married until Mr. Mather's death on April 5, 1951, at age 94.

In the years following Mr. Mather's death, the fortunes of the Cleveland-Cliffs Marine Department began to change. While the company as a whole prospered, it was becoming increasingly difficult to operate a fleet of cargo vessels profitably. The loss of a major shipping contract in 1980 proved disastrous and saw the Cliffs fleet reduced by ten ships, leaving only three active the following year.

At the end of the 1984 shipping season, the remaining ships still in operation were retired and sold, bringing the company's marine operations to a close after more than a century.

A Cleveland-Cliffs captain stands at the bow thruster control in the pilothouse.

The Crew

The traditional Great Lakes shipping season ran from late March to mid-December. This was a period of hectic activity, which saw the ships operating around the clock, seven days a week.

In order to meet this challenge, the crew was organized into watches. The typical crewman worked four hours on, and eight hours off. This schedule ran continuously, and did not provide a day off or vacations during the shipping season.

While the work was hard, employment on a lake freighter provided benefits rarely found on shore. Room and board were free, and the ships were renowned for the quality of the food they served. Also, crewmen who worked for an entire season on board the same ship received a bonus, which was customarily ten percent of the sailor's wages for the year.

The ship's crew consisted of three departments — deck, engineering, and the galley. The deck department, headed by the ship's captain, was responsible for navigation and meeting the business commitments of the ship's owners. The engineering department, led by the chief engineer, was responsible for all physical plant operations of the vessel. The galley department, headed by the steward, was in charge of food service as well as housekeeping and linens.

Deck Department

The deck, or navigation department, lived and worked at the forward end of the ship. Led by the captain, this group included mates, wheelsmen, watchmen, and deck hands. Deckhands were general laborers. They opened and closed cargo hatches, handled mooring lines, and performed a wide variety of maintenance tasks working under supervision of the first mate. Deckhands were entry level employees, meaning that they could begin work without specific technical training. They worked an eight-hour day, with overtime as necessary. On board the *Mather*, they were housed in the dunnage room.

Watchmen served as lookouts assisting the pilothouse crew while maneuvering in narrow or confined waterways and during conditions of poor visibility. The wheelsmen stood four-hour watches steering in the pilothouse. These crewmen lived in shared rooms in the forward deckhouse and were supervised by the ship's mates.

The mates were licensed by the Coast Guard and shared several important tasks. They kept the ship's payroll records, supervised loading and unloading, stood watches in the pilothouse when the ship was underway, conducted lifeboat and other safety drills, and made certain that the vessel's collection of navigation charts was current and complete.

The captain was head of the deck department and was the final authority for all matters relating to the ship. He was also licensed by the Coast Guard and in early years very likely to have risen from the ranks, beginning work as a deckhand and serving in all intermediate positions before commanding a ship. Unlike most other members of the crew, the captain did not stand watches but was always on call. The captain was routinely on duty in the pilothouse during bad weather, poor visibility, docking, casting off, or negotiating rivers. In practice, he was likely to put in more hours than anyone else in the crew.

A deckhand's life was not all hard work. These two relax for a moment on the *Mather*'s spar deck.

Engineering Department

The ship's engineering staff lived and worked in the stern. In the days of coal-fired steamships, the entry-level job in this department was held by the coal passers. These laborers handled coal and performed other tasks under supervision of the firemen.

The firemen actually stoked the fireboxes with coal to generate the heat necessary to create steam in the boilers.

In the engine room itself, the staff consisted of wipers, oilers, watch engineers, and the chief engineer.

Wipers were responsible for cleanliness and order in the engine room. They cleaned, painted, and performed other housekeeping tasks under supervision of the watch engineers.

Oilers spent their four-hour watches making rounds. This consisted of following a route through the engine room to make certain that all the machinery was properly lubricated and operating correctly. Their work included feeling moving parts of reciprocating engines with their bare hands to insure that they were not overheating.

There were three watch engineers, numbered to correspond to the mates. They were known as the first, second, and third assistant engineers. Collectively they were responsible for the engine, boilers, and the deck machinery. They were licensed by the Coast Guard and stood four-hour watches. In addition, many ships carried a day engineer to perform maintenance tasks. All of them were ultimately responsible to the chief engineer.

On board a lake freighter, the chief engineer's authority was exceeded only by the ship's captain. Like the captain, he did not stand a regular watch schedule, but was always on call. The chief was on duty whenever the ship was arriving or departing from a dock or maneuvering in rivers. He also had an office in

The stoves in *Mather*'s galley.

the engine room where he was available to consult with the watch engineers. The chief engineer was also very likely to have risen from the ranks, beginning his career years earlier as a wiper or a coal passer.

Galley Department

The galley department, also located aft, played a critical role in the life of the ship. The crew's meals were prepared by the steward and his staff. The staff included porters and assistant, or second cooks.

The porters were responsible for serving meals and cleaning the galley. When guests were on board, extra porters were assigned to serve their needs. Porters held the entry-level job in the galley department. The next step up was to the position of second cook. These crewmen prepared meals under supervision of the ship's steward.

The steward planned meals, ordered groceries, and was responsible for ensuring that all on board were satisfied with the quality of food provided by the galley.

In order to provide meals for crewmen after the galley's normal working hours, a night kitchen was provided. This included a well-known tradition aboard lake freighters — the night lunch. The night lunch included a selection of the day's leftovers as well as a platter of cold cuts, pies, and pastries. Sailors seldom went hungry on an ore boat.

In more recent years, working conditions on board lake freighters have changed significantly. Work schedules now provide for time off during the shipping season, and many officers receive their training at an academy instead of beginning as deckhands or wipers.

A place setting in the guest dining room. The china is a replica of that in use in the 1940's.

The War Years

Before dawn on September 1, 1939, the German army invaded Poland, beginning the Second World War. This conflict was quick to impact the Great Lakes region. Canada was at war with Germany, and while the United States had a policy of strict neutrality, it was not difficult to see where American sympathies lay. The immediate result was a greatly increased demand for steel.

The shipping industry was quick to respond, and lake freighters and sailors idled by the depression found themselves back in demand. As one of the primary suppliers of iron ore to area steel mills, Cleveland-Cliffs and the *William G. Mather* played a vital role in the Allied war effort from the start. In April 1941, the *Mather* led a group of Cliffs ore boats to Duluth, Minnesota. With the aid of a railroad car ferry acting as an icebreaker, they opened navigation in spite of severe weather and were instrumental in creating a record volume of steel production for the year.

Before the year was out, the Japanese attack on Pearl Harbor drew the United States into the war directly. The ore boats certainly did their part responding to this emergency. While they did not carry guns or wear gray paint, they played a pivotal role. The iron ore they delivered was made into steel that was fabricated into rifles, aircraft engines, tanks, destroyers, and every other type of war material imaginable.

It has often been said that the Axis powers were overwhelmed by American industry, and that industry relied on the steady supply of steel made possible by Great Lakes shipping.

Repowering and Automation

When the *Mather* went into service in the summer of 1925, she was powered by a quadruple expansion reciprocating steam engine. Steam for this power plant was supplied by three coal-fired boilers, which were also hand-fired. This meant that the ship's crew included firemen and coal passers, who worked in four-hour shifts shoveling coal by hand around the clock at a rate of about one and a half tons per hour. This coal was stored on board in a large hopper known as a bunker. This was loaded through a hatch located forward on the boat deck. When the bunker, which contained 400-500 tons was full, the firemen could easily reach the coal in it themselves. As the coal was consumed, it became necessary for the coal passers to enter the bunker with wheelbarrows and bring coal into the firehold so firemen could concentrate on stoking the boilers.

The work did not end there. The fireboxes had to be cleaned out and the ashes disposed of, and all these tasks had to be performed under difficult conditions. The temperature was very hot, and steam pressure had to be maintained regardless of how bad the footing got when the ship was in rough weather.

In the engine room stood the reciprocating steam engine that powered the ship. This was a tall vertical structure with four cylinders mounted fore and aft on the ship's centerline. This general type of engine enjoyed wide popularity on the Great Lakes in this era. They were strong, simple, reliable, and easy to operate and maintain. They were fascinating to watch in operation because of the large scale of their many visible moving parts. An observer could easily see the crankshaft and connecting rods and see the motion of the valve gear operating the intake and exhaust valves. This type of engine was direct drive,

Taken from the present watch engineer's position looking aft, this February 1954 photo shows the new fuel tanks under construction.

meaning that the speed of the engine was the same as the speed of the propeller in the water. At the *Mather's* cruising speed of 12 knots, the 15-foot diameter propeller turned at 90 RPM.

The *Mather's* original power plant produced 2,500 horsepower and was manufactured by the Great Lakes Engineering Works, the ship's builder.

By the early 1950's, all this was no longer state of the art. New technology presented a vast improvement in the steam turbine. The turbine, while still a steam engine, worked on a different principle than the up and down motion of the reciprocating engine. The turbine consists of a horizontal shaft, surrounded by blades like a fan, and housed in a circular casing. High-pressure superheated steam directed across the blades caused the shaft to turn very rapidly. At full throttle, the turbine speed is 7,000 RPM, much too fast for the propeller to work effectively. A set of reduction gears is used to obtain the lower RPM necessary for the propeller shaft.

The *Mather*'s engine consists of two turbines joined together and is known as a cross-compound steam turbine. It was manufactured by the DeLaval Company of Trenton, New Jersey. At 5,000 horsepower, the new engine, installed during the winter of 1953-54, is twice as powerful as the original power plant.

At the same time, the three coal-fired boilers were replaced with a single oil-fired unit. The coal bunker was removed and fuel tanks with a capacity of 110,000 gallons of fuel oil were installed. With the new fuel system, hand firing became a thing of the past, enabling the crew size to be reduced significantly, thereby lowering the ship's operating cost and increasing its profitability.

Ten years later, in 1964, a system of automated controls was installed to monitor operation of the *Mather*'s single boiler. This modification reduced the number of engine room crewmen even further and is one of the earliest applications of a computer in industry. At the

Mather's new cross-compound steam turbine was installed during the winter of 1953-54. At 5,000 horsepower, the new engine is twice as powerful as the original power plant. It was manufactured by the DeLaval Company of Trenton, New Jersey.

same time, a dual-propeller bow thruster, the first of its kind, was installed to significantly reduce the ship's dependence on tugs when maneuvering.

To recognize these innovations, the single boiler installation, the automated control system, and the bow thruster, the *William G. Mather* was declared a National Historic Mechanical Engineering Landmark in 1995.

Mather's main engine control panel.

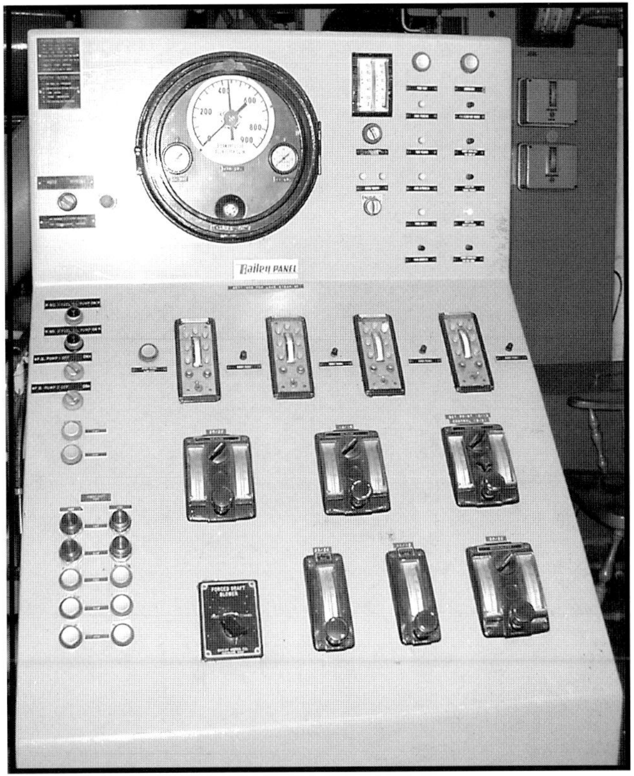

Mather's automated controls for her single boiler.

Part of the repowering process was giving the *Mather* a larger funnel (smokestack). Here during a late March 1954 snow-storm, a crane lowers the ship's newly fabricated funnel into place.

The new funnel is seated with the workers scaffolding still surrounding it.

This excellent photo shows the *Mather* as she appeared after repowering in 1954.

Idled by a steel strike in the summer of 1952, six Cliffs freighters wait at the East 9th Street Pier for the dispute to be settled. The *William G. Mather* is fifth from right.

This view from April 1941 shows the *Mather's* captain, a mate, and a wheelsman on duty in the pilothouse. Notice how sparsely equipped this space is compared to the ship's present configuration.

Nearing retirement after fifty years as a Great Lakes sailor, Captain Charles A. Anderson relaxes for a moment at the *Mather*'s chart table.

During the winter months, freighters went into lay-up. Seen here in the Cuyahoga River in Cleveland, Ohio, Cleveland-Cliffs steamers *Yosemite* and *Grand Island* wait for the coming of spring.

This view shows a brief interlude early in the *Mather*'s career--her work transporting new cars to distribution points around the lakes.

The volume of coal smoke indicates how hard the fireman and coalpassers are working as the *William G. Mather* battles its way through an ice field in April 1941.

The Coast Guard cutter *Tahoma*, one of the loyal companions of the Great Lakes.

The *Sainte Marie* was a railroad car ferry. This type of ship, once a common sight on the Great Lakes, operated year round and was very heavily constructed to cope with winter ice. It was not unusual for them to act as icebreakers when conventional ships needed assistance.

In April 1941, the *Mather* makes slow progress through the ice. The car ferry *Sainte Marie* assists while the Coast Guard cutter *Tahoma* stands by.

This photo, taken by a *Life* magazine photographer from the *Mather*'s Texas deck, shows the ships she led to Duluth to open the 1941 season of navigation.

The *Mather* leads the way as a fleet of bulk freighters struggles to reach Duluth, April 1941.

This photo shows the *Grand Island* being unloaded with a Hulett unloader. Note the rail cars under the loading platform.

With snow on her deck, the *Mather* loads iron ore early in her career.

Technical Data

Name: William G. Mather   *Built:* 1925 by Great Lakes Engineering Works, Ecorse, MI
Hull #: 250   *Type:* Steel bulk carrier, single screw steamer
Dimensions: Length overall 618'0" Length of keel 592'0" Beam 62'0"
Depth 32'0" Midsummer draft 22'0"
Tonnages: Gross Registered 8662 Deadweight 14750
Carrying capacity: Iron ore (long tons) 13,950 Coal (short tons) 14,000
Machinery : Boiler (1954) one water tube, super heater, forced draft. Babcock and Wilcox
Pressure-psi 450 Heating area sq feet 8775
Fuel usage at lake speed: Gallons per hour 391.8 Gallons per mile 26.6
Engine (1954): Geared cross-compound steam turbine. Delaval Steam Turbine Co.
Shaft horse power: 5000 (5500 max) Average lake speed 15.3 MPH

Scale 1 inch = 1050 inches *Drawing by Randall Shoker*

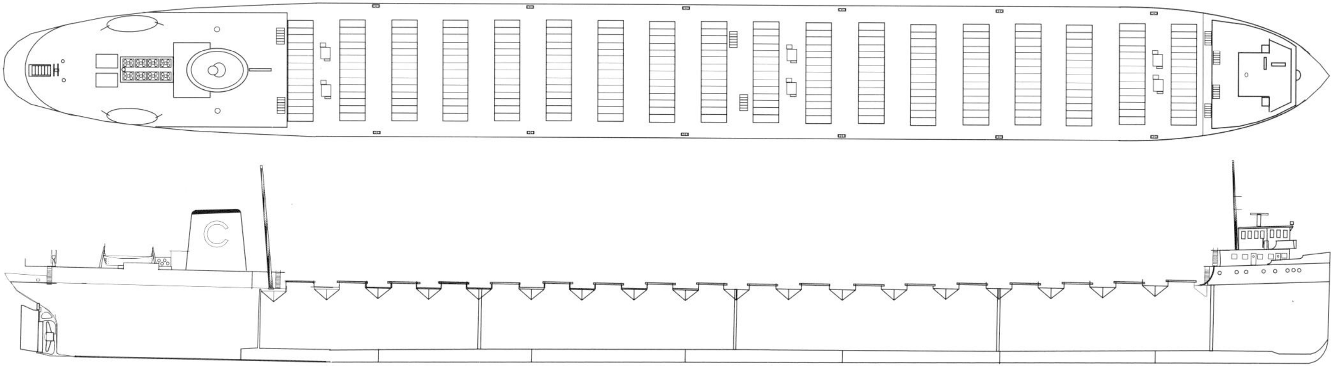

Acknowledgments

I am grateful to Randy Shoker at Oxford Museum Press, and Holly Holcombe, Executive Director of the *William G. Mather* Museum. Both of these good people committed valuable resources to this book at a time when it was little more than an idea in my head.

Also at the *William G. Mather* Museum, Assistant Curator Tracie Drake provided invaluable help locating and verifying historical background information. In addition, longtime museum volunteer Al Zimmie took many of the photos that document the appearance of the ship today. The early black and white photographs appearing in this book are a great help in telling the ship's story. The majority of them were made available through the kindness of Bill Barrow, Special Collections Librarian at Cleveland State University.

I also wish to note the vital, although indirect, contribution of my dear friend Ted Worthington. Had he lived to see it, no one would have been more gratified by the publication of this book.

My mother led me early to a love of books, and my father's lifelong interest in Great Lakes shipping first drew my attention to the subject. Also, the original manuscript of this book benefited greatly from his careful review. As always, his advice was sound.

Thomas G. Matowitz Jr.
Mentor, Ohio
March 2001

Photo Credits

Al Zimmie *Pages 9,23,26,27,29 (photo on right)30,34,36,37,38,39,40,43,44,45,46,55, 56,61,back cover*

Thomas Matowitz Jr. *Pages 8,10,24,25,28, 32,33,41,42,*

William G. Mather Museum *All other photos*

The Steamship *William G. Mather* Museum is permanently moored at the East 9th Street Pier in Downtown Cleveland's North Coast Harbor Park. The Museum is open May – October for touring as follows:

May, September and October:

Fridays and Saturdays, 10 AM – 5 PM
Sundays, 12 PM – 5 PM
Note: Closed Labor Day Weekend due to special programming

June, July and August:

Mondays-Saturdays, 10 AM–5 PM
Sundays, 12 PM – 5 PM

Membership and general information can be obtained by calling (216) 574-9053.

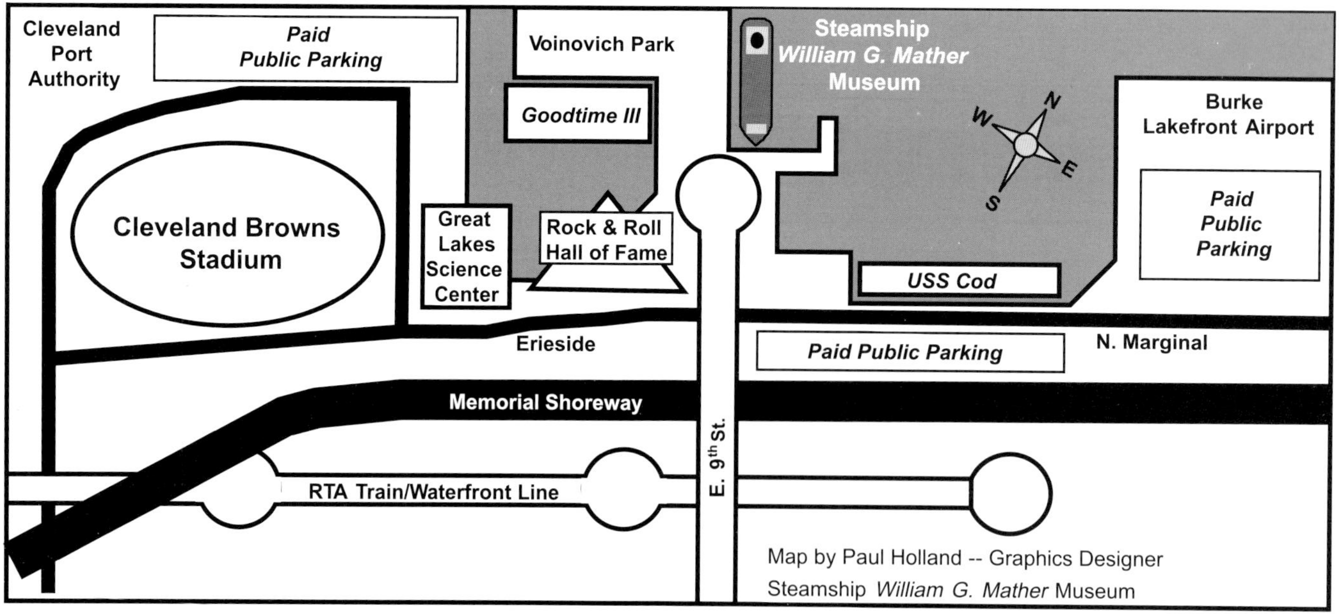